R2R WHITE SERIES

Contents

REAL TO REPLICA WHITE SERIES Volume 3
US Navy & US Marine Corps A-4 Skyhawks

AF874500

REAL

A US Marine Corps A-4M from VMA-331 'Bumblebees' at MCAS Cherry Point
CREDIT: US NAVY

REPLICA

US Air Force F-4 Phantoms

Designed by Andy Folds

ISBN: 978-1-7392977-0-1
Printed and bound in the United Kingdom

PHOENIX
SCALE PUBLICATIONS

Acknowledgments
Thanks are due to the following for their help in the preparation of this title. US Navy Public Affairs, US Marine Corps Public Affairs, McDonnell Douglas, Luc Colin, Brian Wakeman, Jezz Coleman, Andy Renshaw, Hans Spruijt and Mario Serelle.

INTRODUCTION

US Navy & US Marine Corps Skyhawks

The Douglas A-4 Skyhawk is a single-seat subsonic carrier-capable light attack aircraft developed for the United States Navy and United States Marine Corps in the early 1950s. The delta-winged, single turbojet engined aircraft was designed and produced by Douglas Aircraft Company, and later by McDonnell Douglas. It was originally designated A4D under the US Navy's pre-1962 designation system, and was designed as a relatively light aircraft, with a maximum take-off weight of 24,500 pounds and a top speed of 670 miles per hour. The aircraft's five hardpoints support a variety of missiles, bombs, and other munitions, and was capable of carrying a bomb load equivalent to that of a World War II–era Boeing B-17 and could deliver nuclear weapons using a low-altitude bombing system and a loft delivery technique. The A-4 was originally powered by the Wright J65 turbojet engine; but from the A-4E onwards, the Pratt & Whitney J52engine was used. Skyhawks played key roles in the Vietnam War, the Yom Kippur War, and the Falklands War

The Skyhawk was designed by Douglas Aircraft's Ed Heinemann in response to a US Navy call for a jet-powered attack aircraft to replace the older Douglas A-1 Skyraider. Heinemann opted for a design that would minimise its size, weight, and complexity. The result was an aircraft that weighed only half of the Navy's weight specification. It had a wing so compact that it did not need to be folded for carrier stowage, and the diminutive XA4D-1 prototype set a world speed record of 695.163mph on 15 October 1955. Due to its small size, it could be operated from the older, smaller World War II-era aircraft carriers, as these older ships were often unable to accommodate newer Navy fighters such as the F-4 Phantom II and F-8 Crusader, which were faster and more capable than the A-4, but significantly larger and heavier. The US Navy operated the A-4 in both regular Navy and Naval Reserve light attack squadrons (VA). Although the A-4's use as a training and adversary aircraft would continue well into the 1990s, the Navy began removing the aircraft from its frontline attack squadrons in 1967, with the last ones being the Super Foxes of VA-55/212/164, retiring in 1976.

The Marine Corps would not take the Skyhawk's replacement, the A-7 Corsair instead they kept their Skyhawks in service and ordered a new A-4M

An A-4C from VA-93 'Blue Blazers' is positioned for launch aboard the USS Ranger in March 1965
CREDIT: US Navy

An A-4E from VA-94 'Mighty Shrikes' overflies the USS Bon Homme Richard
CREDIT: US NAVY

A US Marine Corps A-4M from VMA-331 'Bumblebees' at MCAS Cheery Point
CREDIT: US NAVY

model. The last USMC Skyhawk was delivered in 1979, and they were used until the mid-1980s before they were replaced by the equally small, but more versatile STOVL AV-8 Harrier, and VMA-131, 'Diamondbacks' retired its last four OA-4Ms Skyhawks on 22 June 1994. Trainer and single seat versions of the Skyhawk remained in Navy service, however, finding a new lease on life with the advent of adversary training, where the nimble A-4 was used as a stand-in for the MiG-17 in dissimilar air combat training (DACT). It served in that role at Topgun until 1999.

The A-4's nimble performance also made it suitable to replace the F-4 Phantom when the Navy downsized its aircraft for the Blue Angels demonstration team, until F/A-18 Hornets were available in the 1980s. The last Navy Skyhawks, TA-4J models belonging to the composite squadron VC-8, remained in military use for target towing, and as adversary aircraft, for combat training at Naval Station Roosevelt Roads. These aircraft were officially retired on 3 May 2003.

REAL
TO REPLICA

CHAPTER 1

Design and Definition

Seen at the types roll out, Ed Heinemann stands on the access ladder of the first XA4D-1

The original windscreen of the XA4D-1 was frameless

The US Navy was becoming increasingly alarmed at the growing trend towards the increasing weight in fighters and attack aircraft of the day such as the USAF's F-86 Sabre and the Navy's F9F Panther, and Douglas came up with a rather daring proposal for a jet fighter weighing only 7000 pounds, a design which eventually led to the A-4 Skyhawk. The Navy showed some interest, but since they were already involved in the consideration of several other fighter designs, they suggested that Douglas should apply the same sort of philosophy to the design of a carrier-based attack aircraft. This aircraft would be intended for the nuclear strike role, with a top speed of 500mph, a combat radius of 345 miles, a 2000lb weapons load, and a maximum gross weight of less than 30,000 pounds. A Douglas a team led by Ed Heinemann responded a couple of weeks later with a proposal that exceeded these requirements by a great margin. The normal loaded weight of the aircraft would be only 12,000lbs, less than half the limit specified by the Navy, and the top speed was 100mph more and the combat radius 115 miles greater. Douglas was authorised to proceed with further design studies. During the evaluation, the range requirements were increased, raising the gross weight to 14,000lbs

The design team came up with a low-winged jet-powered aircraft with a modified delta planform, with a quarter chord sweep of thirty-three degrees. The span was only twenty-seven feet six-inches, which eliminated any need for wing folding and saving a lot of weight and complexity. The wing had three one-piece spars with span-wise stiffened skin. The delta shaped wing formed a single box with integral fuel tankage, and the upper and lower skins were single pieces, and the spars and stringers were continuous from tip to tip. The wing leading edge was equipped with automatic leading edge slats, and split flaps were provided on the trailing edge. The leading edge were designed to drop automatically at the appropriate speed by gravity and air pressure, thus saving weight and space by omitting actuation motors and switches. Most of the wing between the spars contained an integral fuel tank with a capacity of 560 gallons. The aircraft had a normal tail, with a rudder and a set of elevators, and the dorsal fin had a delta shape, and had a rudder set at its rear. The rudder, originally traditionally faired over was later uniquely constructed as a single panel, reinforced with external ribs to save weight.

The horizontal tail plane was set at the lower part of the vertical tail, just above the tailpipe. The horizontal stabiliser was electrically adjustable in incidence, and could be adjusted for trim throughout the entire flight range, and a large speed brake was provided on each side of the rear fuselage. The engine was to be, built by Curtiss-Wright under the designation J65. The A-4 was originally powered by a Wright J65 turbojet engine, a licence-built version of the British Armstrong Siddeley Sapphire turbojet but from the A-4E onwards, a Pratt & Whitney J52 engine was used. The engine was mounted in the fuselage centre with air intakes mounted high on both sides of the fuselage aft of the cockpit, with a single exhaust in the tail. The engine was accessed for service or replacement by removing the aft section of the fuselage and sliding out the engine. This obviated the need for access doors with their hinges and latches further reducing weight and complexity. The internal fuel capacity was 770 US gallons, carried in integral wing tanks and in a self-sealing cell aft of the cockpit and between the engine air ducts. The A-4 pioneered the concept of 'buddy' air-to-air refuelling. This allows the aircraft to supply others of the same type, reducing the need for dedicated tanker aircraft. This allowed for greatly improved operational flexibility and reassurance against the loss or malfunction of tanker aircraft, though this procedure reduces the effective combat force on board the carrier. A designated supply A-4 would mount a centre-mounted 'buddy store', a large external fuel tank with a hose reel in the aft section and an extensible drogue refuelling basket. This aircraft was fuelled up without armament and launched first. Attack aircraft were then armed to the maximum and were given as much fuel as was allowable by maximum take-off weight limits, which was far less than a full tank. Once airborne, they topped off their fuel tanks from the tanker using the A-4's original fixed, later angled refuelling probe on the starboard side of the aircraft nose. They could then sortie with both full armament and fuel loads. The A-4 was rarely used for refuelling in US service after the KA-3 Skywarrior became available aboard the larger carriers.

All the offensive weapons were to be carried externally on three stations, one underneath the

XA4D-1 flew for the first time from Edwards AFB on June 22, 1954

The XAD-1 originally had no arrestor hook fitted

A full scale mock-up of the Skyhawk design from 1952

fuselage centreline and one underneath each wing, later two. The internal armament was to be a pair of 20mm cannons, one in each wing with 100 rounds per gun (the A-4M and types based on the A-4M had 200 rounds per gun), plus a large variety of bombs, rockets, and missiles could be fitted, with the ability to carry a single Mk.12 nuclear weapon. The combat radius with this weapon and internal fuel only was 400 miles. The aircraft was also designed to be able to make an emergency landing, in the event of a hydraulic failure, on the two drop tanks nearly always carried by these aircraft. Such landings resulted in only minor damage to the nose of the aircraft which could be repaired in less than an hour.

The tall main undercarriage members were attached to the inner wing trailing edge, and retracted forward and rotated through ninety-degrees to fit into wells in the leading edge of the wing. The wing was sufficiently thin that long fairings had to be fitted underneath the wing to cover the landing gear legs when retracted. The nose landing gear retracted forward into a well in the nose, and the forward-retracting landing gear had the advantage that in that emergency extension systems were not required, since the airstream flow will lock the gear down after free fall. The A-4 was also designed to be able to make an emergency landing, in the event of a hydraulic failure, on the two drop tanks nearly always carried by these aircraft. Such landings resulted in only minor damage to the nose of the aircraft which could be repaired in less than an hour. The cockpit canopy was of the 'clamshell' variety, opening via a hinge located immediately to the rear, and an upward firing ejector seat was to be provided for the pilot, as was, in early aircraft a nuclear flash shield.

A preliminary mock-up inspection took place in February of 1952 Douglas was given a contract for one aircraft On June 12, 1952, with the designation was XA4D-1, and the BuNo of 137812. The project was financed by diverting funds from the cancelled A2D Skyshark program, and a final mock-up inspection took place in October of 1952. By this time, the Navy had ordered nine production aircraft, which was soon increased to nineteen. The first XA4D-1 was assembled at the Douglas El Segundo plant, and virtually 'hand-built' and was rolled out of the factory in February of 1954, the aircraft being given the popular name Skyhawk. In press releases, the plane was often referred to as 'Heinemann's Hot Rod', but soon gained the other nicknames of 'Scooter', 'Bantam

The second production A4D-1

The XAD-1 at Edwards AFB in 1954

Preparing the XA4D-1 for a test flight

Bomber', Tinker Toy Bomber' and 'Kiddiecar'. The windscreen of the cockpit was frameless, and the nose was provided with a long instrumentation probe, and the pilot sat on a NAMC Type II ejection seat. The original ejector seat was re-designed by Douglas from one from the F4D Skyray and only used eighty parts instead of 240 weighed just 40lbs.

Only the centreline weapons pylon was fitted, there was no carrier arrester hook, and no armament was fitted. The XA4D-1 was trucked out to Edwards AFB, and its first flight was delayed by the late delivery of its Curtiss-Wright J65-W-2 turbojet. So, the aircraft finally took to the air at Edwards AFB on June 22, 1954, with test pilot Robert Rahn being at the controls. Later the XA4D-1 was fitted with most of the features of the production A4D-1, including a tail hook, a jet pipe fairing, vortex generators, and all three weapons pylons. Nine of the first nineteen Skyhawks were designated YA4D-1 for flight testing purposes, and the first factory-to-fleet delivery of an A4D-1 was made to VA-72 at NAS Quonset Point, on 26 September 1956. Shortly thereafter, the first Pacific fleet Skyhawk deliveries were made to VF(AW)-3 at NAS Moffett Field, CA, and VMA-224 at MCAS El Toro, CA. In all, eighteen Navy and Marine squadrons took delivery of the A4D-1 (A-4A). The Skyhawk remained in production until 1979, with 2,960 aircraft built, including 555 two-seat versions, and the last production A-4, an A-4M for VMA-331 had the flags of all nations that operated the A-4 painted on its fuselage sides.

The XAD-1 was literally 'hand built'

Undertaking refuelling trials

CHAPTER 2

Skyhawks by Variant

A4D-1 Skyhawk BuNo 139966 of Attack Squadron VA-43 'Challengers' on the deck of the training carrier USS Antietam
CREDIT: US Navy

The A4D-1/A4A Skyhawk

The A4D-1 was the first production version of the Skyhawk, and BuNo 137813 took-off on its first flight on August 14, 1954, only two months after the first flight of the prototype XA4D-1. A total of 165 A4D-1s were built, the last one being delivered in 1957. Of these, A4D-1 BuNo137815 crashed on a test flight on January 31, 1955, killing Douglas test pilot James Verdin, A4D-1 BuNo137816 was used as the carrier suitability test aircraft, A4D-1 BuNo 137817 was the weapons systems test aircraft, A4D-1 BuNo137820 was used for bomb testing, A4D-1 BuNo137821 was used for locked slats and wing fence testing, and A4D-1 BuNo137824 was used for electronics testing. The A4D-1 was much the same as the XA4D-1, but had a 'sugar scoop-shaped' jet tailpipe fairing, and was fitted with an arrester hook. Three weapons pylons were fitted, one underneath the fuselage centreline and one underneath each wing, just outboard of the main landing gear, allowing 5000lbs of ordnance to be carried. Alternatively, up to three drop tanks could be installed, with a combined capacity of 800 gallons. A large blade antenna for the UHF radio was placed immediately aft of the cockpit, and the frameless windshield of the XA4D-1 was replaced by a framed version. An internal armament of two 20mm Colt Mk.12 cannons was also fitted, with one gun in each wing root, carrying 100 rounds per gun. The aircraft did not carry any radar, the nose cone being

A4D-1 Skyhawk (BuNo 139944) of VA-72 'Hawks' during carrier suitability tests
CREDIT: US Navy

A4D-1 Skyhawk BuNo 139966 from VA-34 'Blue Blasters'
CREDIT: US Navy

An A4D-1 Skyhawk from VA-153 'Blue Tail Flies' on the starboard catapult of the USS Hancock
CREDIT: US Navy

filled with avionics equipment, and was powered by a Wright J65-W-4 or W-4B turbojet. This was the only variant with a conventionally skinned rudder, and no refuelling probe. Electronics included and AN/ASQ-17 group consisting of UHF communications, IFF, Coder, and UHF Direction Finding functions. Externally and AN/ARN-12 Marker-Beacon Receiver and an AN/.ARN-14 VOP Receiver could also be carried. On October 15, 1955, Lt Gordon Gray set a new 100km closed circuit speed record of 695.163mph in A4D-1 BuNo 137820, taking the record away from the USAF's F-86H, and was the first 'attack' aircraft to hold this record. In October 1956, the first A4D-1 was taken on strength by VA-72, and the first West Coast fleet squadron to receive the type was VA-93, who then deployed to the Western Pacific aboard the USS Ticonderoga in September of 1957. The first Marine Corps squadron to receive the Skyhawk was VMA-224, who took their first A4D-1 in January of

An A4D-1 from VA-44 'Hornets'. Note that this aircraft is not fitted with the 20mm Mk.12 cannon, and the weapons load consists mostly of conventional iron and nuclear bombs
CREDIT: US Navy

Seen at NAS Alameda an A4D-1 with both Navy and Marines markings
CREDIT: US Navy

1957. Eventually, some eighteen Navy and Marine Corps attack squadrons were issued with A4D-1s, and almost all of these squadrons were former Navy VF or USMC VMF jet fighter units that were then redesignated as attack VA or VMA squadrons. The Navy Skyhawk squadrons served aboard carriers alongside propeller-driven Skyraider, and as the A4D-1 was replaced by later versions of the Skyhawk, they were transferred to Navy and Marine Corps Reserve squadrons. In September of 1962, the A4D-1 was redesignated A-4A to reflect the fact that later versions of the Skyhawk were much more capable, and to reduce the effective number of aircraft in the attack inventory, in an effort to procure more aircraft for the Vietnam war. A total of 166 aircraft were built.

A4D-1s from VA-113 'Stingers' lined up at NAS Miramar in August 1957
CREDIT: US Navy

A4D-2/A-4B Skyhawk

The A4D-2 was the second production version of the Skyhawk, and differed internally from the A4D-1 by having a strengthened rear fuselage structure. The A4D-2 also featured a new single-skin rudder, which was built 'inside-out' with the ribs showing, and

An A4D-1 Skyhawk of VA-72 'Hawks' is loaded aboard the USS Randolph in 1956

An A4D-2 Skyhawk of VMA-311 'Tomcats on' the ground at MCAS El Toro *CREDIT: US Navy*

was bonded together rather than being riveted. This also corrected a tendency of the rudder to vibrate at high speed, thus creating fatigue cracks, and was also provided with an external trim tab. Perhaps the most noticeable external change was the addition of a non-retractable in-flight refuelling probe to the starboard side of the fuselage, and in addition, the centreline attachment point was given the ability to carry a 300 US-gallon 'buddy' refuelling store. This pod had a built-in drogue and hose, which could be used to refuel other aircraft in flight. Internally, additional navigational equipment was provided, and the weapons delivery system was improved with the capability to carry and launch the Bullpup air-to-ground guided missile. Additional improvements were also made to the flight control systems, the landing gear was strengthened and a pressure fuelling system was added, and and improved gunsight and bombing system improved ordnance accuracy. Also, the pilot now sat on an Escapac 1 rocket-boosted ejector seat, and the A4D-2 was equipped with a radar altimeter, with the transmitter/receiver unit mounted in a canoe shaped housing under the left wing near the wing tip. The aircraft was originally powered by a Curtiss-Wright J65-W-16A, whilst some were upgraded to a newer version of this engine, the J65-W-20, and the A4D-2 was equipped with a radar altimeter, with the transmitter/ receiver unit mounted in a canoe shaped housing under the left wing near the wing tip. Improved electronics

An A4D-2 Skyhawk of VA-72 'Blue Hawks' prepares to launch *CREDIT: US Navy*

Only a centreline weapons pylon was fitted, there was no arrester hook *CREDIT: Douglas Archives*

An A4D-2 Skyhawk of VA-34 'Blue Blasters' armed with a Mk.7 nuclear weapon prepares to launch from the USS Saratoga *CREDIT: US Navy*

A4D-2 Skyhawk 142134 from VA-76 'Spirits' aboard the USS Intrepid *CREDIT: US Navy*

A4D-2 Skyhawks of VA-86 'Sidewinders' in formation *CREDIT: US Navy*

An A4D-2 Skyhawk from VA-81 'Crusaders', Det.45 about to launch from the USS Essex *CREDIT: US Navy*

included the AN/ASQ-17 Electronic Central Control System. The hydraulic system was improved, and a single point refuelling system added. The first A4D-2 flight occurred on March 26, 1956, flown by Douglas test pilot Dru Wood, and the first A4D-2 was delivered to the fleet in September of 1957, with the first squadron to receive the type being the Marine Corps squadron VMA-211, with VA-12 being the first Navy unit to receive the type in February 1958. The A4D-2 also flew the first Skyhawk missions where there was a real possibility of combat, when VA-83 covered the Marine landings in Lebanon in 1958. In all some 542 A4D-2s were built. The A4D-2 was redesignated A-4B in September of 1962, and as with the A-4A, when later Skyhawk models became available, many A-4Bs were relegated to Navy and Marine Corps reserve units, and several surplus A-4Bs were later refurbished and sold to foreign air forces. When delivery to fleet units of later versions of the Skyhawk began, some A4D-2 airframes were assigned to Naval Reserve squadrons and re-designated TA-4B, although no structural modifications were involved. During the latter part of the 1950s and early 1960s the A4D Skyhawks played a major part in nuclear weapons delivery from carrier based aircraft, and each Skyhawk pilot was assigned one or more targets at the beginning of his deployment.

An A-4B from the Naval Missile Centre, with three AQM-37A target drones *CREDIT: US Navy*

An A4D-2 from VA-172 'Blue Bolts' won the deck of the USS Franklin Roosevelt in 1959

An A-4C Skyhawk from VA-64 'Black Lancers' is readied for launch aboard the USS America during carrier qualifications
CREDIT: US Navy

The A4D-2N/A-4C Skyhawk

All earlier Skyhawk variants were basically day-only attack aircraft, and did not have the capability to operate at night or in bad weather. So, the next Skyhawk variant, the A4D-2N, was intended to give the aircraft an all-weather capability. As such, the A4D-2N, the third and most numerous Skyhawk variant, and was equipped with an AN/APG-53A terrain-clearance radar housed in a nine-inch longer nose, an autopilot, an all-attitude indicating gyro, an improved Escapac 1A-1 low-level ejector seat, a windshield wiper and a revised cockpit layout. The AN/ASQ-17 Electronic Central Control System was retained. The 'N' designation represented the 'night' capability delivered by the addition of a radar. The A4D-2N was initially powered by a J65-W-16A turbojet, but many later received the J65-W-20. The A4D-2N flew for the first time on August 21, 1958, and became operational in February of 1960, with the VMA-225, and with VA-192 in May 1960. The A4D-2N was redesignated A-4C in September of 1962, although at the time the A4D-2N was still in production. During the production run of the A-4C, the production line was moved from the El Segundo plant to the Long Beach plant, with final assembly being done at a

-4C from VMA-324 'Vagabonds' demonstrates the JATO technique
REDIT: US Navy

Checking over the flight manuals for this A-4C
CREDIT: US Navy

An A-4C from VA-34 'Blue Blasters' is marshalled into position aboard the USS Saratoga *CREDIT: US Navy*

government-owned plant in Palmdale, California. The A4D-2N was equipped with a small, simple APG-53 radar developed specifically for the Skyhawk as well as a TPQ-10 blind bombing system, an AJB-3 low altitude bombing system and all all-reference autopilot and airstream direction detector. Some 638 A4D-2Ns were built, and some 100 A4D-2Ns that were being phased out of fleet squadron service were modified as A-4Ls to fulfil a requirement for the Navy and Marine Corps Reserves. As the A-4C and A-4L became surplus to Navy and Marine Corps requirements, several of them were delivered to foreign air forces such as Argentina, Singapore, and Malaysia, as can be seen in Chapter 6. Two A4D-2Ns, BuNos 148490 and 148483 were modified in 1961 for evaluation by the US Army as a ground support aircraft. Each of these aircraft was fitted with a drag chute stored in a canister underneath the rear fuselage, and was fitted with low-pressure twin-wheel main undercarriage that retracted into enlarged fairings underneath the wings. The two A4D-2Ns were evaluated by the Army at Fort Rucker, Alabama

Two A4D-2Ns were modified in 1961 for evaluation by the US Army as a ground support aircraft *CREDIT: Douglas Aircraft Corporation*

An A-4C hangs from the catwalk after a deck accident *CREDIT: US Navy*

An A-4C from VA-192 'Golden Dragons' lands aboard the USS Bennington in 1962 *CREDIT: US Navy*

in competition with the Northrop N-156 and the Fiat G.91R. However, it was decided that the Army would not be permitted to operate fixed-wing attack aircraft, leaving responsibility for close air support of ground troops to the Air Force, the Navy, or the Marine Corps. As a result, none of the competitors was ordered by the Army, and the two A4D-2Ns were then converted back to standard configuration and delivered to the Navy.

A4D-5/A-4E Skyhawk

The A4D-5 was a considerably revised and more advanced version of the Skyhawk, and was powered by a Pratt & Whitney J52-P-6A engine, which had a twenty-seven percent lower fuel consumption than J65 of earlier Skyhawk versions, which substantially increased the range. The J52 engine however, required a redesign of the centre fuselage and inlet ducting, with a splitter plate added to the edge of the inner intake next to the wall of the fuselage. Earlier Skyhawks had been designed primarily with the nuclear mission in mind, however, as the nuclear mission was now becoming less important, the A4D-5 was an attempt to adapt the basic Skyhawk design to deliver conventional munitions. The A4D-5 had

An A-4C from VA-172 'Blue Bolts with 'buddy' refuelling pod prepares to launch from the USS Franklin D Roosevelt in 1966 *CREDIT: US Navy*

In 1971 A-4E Skyhawks from VMA-311 'Tomcats' take on fuel at Da Nang, South Vietnam *CREDIT: US Marine Corps*

A great shot of an A-4E from VA-45 Det.1 'Blackbirds' in flight *CREDIT: US Navy*

A-4E 152048, Lady Jessie, from VA-164 'Ghost Riders' *CREDIT: US Navy*

an extra pair of underwing weapons hardpoints, one underneath each outer wing, and these extra stations brought the maximum weapons load to 8200lbs, and allowed for a much wider variety of weapons and fuel options. The nose was lengthened by fourteen-inches to accommodate an AN/ASN-19A navigation computer, and new-equipment including a TACAN, Doppler navigation, a Mk.9 toss bombing system, radio altimeter, and the AJB-3A low-altitude bombing system was installed. On July 30, 1959, approval was given to deliver two A4D-2Ns BuNo 148613 and BuNo 148614 as A4D-5s and the first of these flew on July 12, 1961. The A4D-5 was redesignated A-4E in September of 1962 (the designation A-4D

An A-4E Skyhawk from VA-45 Det.1 'Blackbirds' aboard the USS Intrepid *CREDIT: US Navy*

An A-4E from VMA-131 'Diamondbacks' assigned to the Marine Air Training Detachment at NAS Willow Grove unleaded ordnance in a training mission

An A-4E of VF-101 'Grim Reapers' Key West Detachment

A US Marine Corps A-4E of VMA-322 'Fighting Gamecocks' at NA Moffett Field

was skipped to avoid confusion with the older A4D designation). As such the new A-4E began to enter service in late 1962, with the first squadron to receive the type being VA-23 at NAS Lemoore, California, who received their first A-4Es in December of 1962. A total of 500 A-4Es were built. During its service life the A-4E was retrofitted with the humpbacked avionics compartment immediately behind the cockpit that had been introduced on the A-4F. In addition, many received the more powerful J52-P-8 engine, and some A-4Es had their straight refuelling probe replaced by the canted example that was fitted to the A-4M.

Twenty-eight A-4Es were transferred to Israel in October of 1973 as an emergency measure to make up for the high losses of Israeli Skyhawks during the Yom Kippur War.

An A-4E of the Naval Air Test Center at Patuxent River seen here in August 1967

An A-4E from VA-23 'Black Knights' readies to launch from the USS Midway in February 1964

Assigned to VA-55 'Warhorses' this A-4E has bombs and Bullpup missiles fitted

A glossy A-4F Skyhawk of VMA-142 'Flying Gators' *CREDIT: US Navy*

An A-4F from VA-144 'Roadrunners' in flight *CREDIT: US Navy*

Douglas A-4F Skyhawk

Had it not been for the Vietnam war, the A-4E would have been the last version of the Skyhawk to be produced for the Navy. But after the war began, the A-4F version was designed specifically for use in the war, and was only assigned to Pacific coast squadrons. Thus, the A-4F was the last attack Skyhawk version to be built for the Navy, and it was also the first Skyhawk version to be produced entirely after the introduction of the new unified designation scheme of September 1962, and as such never had an 'A4D' designation. The A-4F was initially ordered in 1965 to make up for the heavy losses that Navy and Marine Corps Skyhawk units had suffered in Vietnam, and was based on the earlier A-4E, but differed in incorporating the nose wheel steering, upper wing lift spoilers, the Escapac 1C-3 ejector and the Pratt & Whitney J52-P-8A engine that had been introduced on the TA-4F. The last A-4E, BuNo 152101 was modified as the first A-4F and flew for the first time on August 31, 1966, and production aircraft were initially delivered to VA-23 and VA-93, which deployed to Southeast Asia in late 1967. A total of 147 A-4Fs were built. During service, the A-4F was retrofitted with a humpbacked avionics compartment located on top of the fuselage immediately behind the cockpit. This compartment was provided for the installation of new types of special electronics equipment dictated by the requirements of the Vietnam War. This rather odd location was selected because there was no more room left inside the Skyhawk fuselage for the installation of any more electronic equipment! Kits were supplied for aircraft already in the field, and those aircraft still under construction at the factory were modified at the production line. This humpbacked pod was also retrofitted to all operational A-4E aircraft and to some A-4C aircraft as well. Some late A-4Fs were fitted with an ALR-45 at the top of the vertical fin, and in addition, a small chin-type DECM antenna was installed underneath the

A line-up of A-4F's from VA-92 *CREDIT: US Navy*

Six Marine Corps A-4Fs from VMA-133 'Dragons' with an EA6A Intruder *CREDIT: US Navy*

nose. Douglas supplied 100 kits to enable the Navy's repair facilities to replace the J52-P-8A engine with the more powerful J52-P-408, and the installation of this engine resulted in a marked improvement in performance. The A-4F also saw service on the 'adversary' role, being stripped of as much equipment as possible the 'Super Fox' made a great dogfighter.

An A-4F from VC-7 'Redtails' in the 1970s *CREDIT: US Navy*

The A-4F with the Blue Angels

A bad accident with the Phantoms of the US Navy's Blue Angels flight demonstration team on July 26, 1973 caused the cancellation of the rest of their season. In addition, over the next few months an energy crisis had appeared in the United States, caused by an oil embargo arising out of the Yom Kippur War. This resulted in gasoline shortages, and as a result, the Navy decided that it would be unwise from a public relations standpoint to fly expensive, fuel-guzzling aircraft in airshows, so they went shopping for a less expensive, more fuel-efficient aircraft for the Blue Angels. The team really wanted the F-14 Tomcat, but it was ruled out on the basis of cost, and the A-7 Corsair was also considered, but was ruled as the fleet had none to spare. So, Navy decided to adopt a reconfigured A-4F. Eight A-4F airframes were selected for initial modification and these had their J52-P-8 engines replaced by the J52-P-408, their guns and armour plating were removed, the avionics package was deleted, and a drogue chute was added. Load feel

Eight A-4F airframes were selected for initial 'Blue Angels' modification *CREDIT: US Navy*

A 'Blue Angel' takes off *CREDIT: US Navy*

A USMC A-4L of VMA-142. This aircraft was later converted to an A-4PTM *CREDIT: US Navy*

bungee-cords were added to provide additional nose-down stick forces during demonstration flights, and a smoke system was fitted, along with a foldable ladder for crew entry in the left gun position and additional ballast was added in the nose. Fearful that they might open asymmetrically during close formation flying in tight turns, the wing leading-edge slats were bolted shut. Over the next few years, additional A-4Fs were modified for the team, and a total of eighteen different Skyhawks served with the Blue Angels, and a TA-4J two-seater was also assigned to the team, and three other two seaters were used temporarily while the prime aircraft was under maintenance. On July 13, 1985, two Blue Angels Skyhawks collided in mid-air during an airshow at the Niagara Falls Airport, and saw to accelerate the changeover to the F/A-18 Hornet, which took place during the winter training ready for the 1986 season. As they completed their tours, some of the Blue Angels Skyhawks were reassigned to adversary squadrons, and the trainer was reassigned to Training Command, whilst others were donated to museums. There are, however, several A-4s in museums painted to represent Blue Angels Skyhawks, even though they never served with the unit. In addition, re-engined A-4Fs were issued to adversarial flight teams as well as to some US Marine Corps reserve units.

A-4L Skyhawk

The A-4L designation was assigned to 100 Naval Air Reserve A-4Cs that were upgraded during the Vietnam War. The upgrade was designed to bring them up to the standards of Skyhawks that were currently in fleet service, in the expectation that there could be two Reserve carrier air wings that would be fully combat-ready should the need arise.

The upgrade included the installation of an uprated Pratt Whitney J-52-P8A, 8B engine, plus the addition of the dorsal avionics pod. In addition, the wing lift spoilers that were first introduced on the TA-4F were included in the modification kit. However, the number of weapons pylons remained at three. The

A-4L Skyhawks from VA-205 'Green Falcons, VA-204 'River Rattlers' and VA-203 'Dolphins' aboard the USS John F Kennedy in August 1971 *CREDIT: US Navy*

An A-4L of VA-205 'Green Falcons' *CREDIT: US Navy*

An A-4M of VMA-211 at MCAS El Toro *CREDIT: US Navy*

first A-4L upgrade was flown for the first time on August 21, 1969, and was the first of 100 aircraft to be so modified. Surplus A-4L aircraft were later reworked to A-4PTMs for delivery to Malaysia.

Douglas A-4M Skyhawk

The A-4M was a version of the Skyhawk specifically designed to fulfil a Marine Corps requirement for an attack aircraft that could fly close air support from short air strips close to the front lines, and was selected over the A-7 Corsair. The A-4M was powered by a new version of the J52 engine, the J52-P-408, which was twenty-percent more powerful than the earlier J52-P-8A, and this higher thrust markedly improved the short-field performance. The new engine was only one percent heavier, and had no increase in specific fuel consumption, and in addition, engine components were installed to reduce the amount of visible engine exhaust. In addition, a self-contained engine starter was provided, and the electrical power generating capacity was increased by sixty-percent.

The A-4M also featured a larger cockpit canopy to improve pilot visibility, especially toward the rear, and a windshield like that used on the trainer version was installed, and the canopy was three-inches wider. A ribbon-type drag chute was installed underneath the engine tailpipe, which helped to slow the aircraft after touchdown, making it possible for the aircraft to operate from short airstrips. The A-4M was also fitted with a revised refuelling probe which canted out to starboard to prevent interference to the wide-angle target acquisition system. A new squared-off tip to the vertical tail was introduced, housing an IFF antenna. Authority to proceed with the A-4M was granted in May of 1969, and two A-4Fs BuNos 155045 and 155049 were reworked into the A-4M configuration to serve as prototypes. The first aircraft

The A-4M was a version of the Skyhawk specifically designed to fulfil a Marine Corps requirement *CREDIT: US Navy*

An A-4M from VMA-214 'Blacksheep' and note the ARBS in the nose *CREDIT: US Navy*

A quartet of 'Black Sheep' from VMA-214 *CREDIT: US Navy*

An A-4M from VMA-234 'Vagabonds' launches a Zuni rocket *CREDIT: US Navy*

was flown on April 10, 1970 by test pilot Walt Smith. Douglas also gave the name Skyhawk II to the A-4M, with fleet deliveries beginning on February 26, 1971. The first Marine Corps unit to receive the A-4M was VMA-324 at MCAS Beaufort, South Carolina, and by 1976, all five active USMC light attack squadrons were using the A-4M. During the ten-year production run of the A-4M, significant improvements were made. Amongst these was a new heads-up cockpit display, an improved weapons display delivery system integrated with an Elliott heads-up display which included both air-to-air and air-to-ground modes. A Hughes Angle/Rate Bombing System (ARBS) was also installed, and featured both television and laser tracking modes, and the seeker for this system was installed in the extreme nose. The sensor package included a television camera and a laser spot tracing system to provide for acquisition and tracking of laser-designated targets. On either side of the nose sensors were antenna for the ALR-45 radar warning system, and another antenna for the ALR-45 radar homing and warning system was fitted to the top of the vertical fin. Below the nose was an antenna for an ALQ-126 deception jammer transmitter and receiver system. For a while, the Navy had considered re-designating the modified A-4M as A-4Y, but this was not adopted. A total of 160 A-4Ms were built including the two revised A-4Fs, and the final aircraft was delivered on February 27, 1979, bringing Skyhawk production to an end. At the time, this was the longest production run for any American tactical aircraft, some twenty-seven years. The A-4Ms served with active-duty Marine Corps squadrons until February 27, 1990, when VMA-211 transferred its last A-4M to MAG-42, a reserve unit at NAS Alameda, California. This ended the operational career of the Skyhawk as an attack aircraft with Marine Corps and Navy regular squadrons. The A-4M continued to serve with Reserve units until 1994, when VMA-131 finally relinquished the last of its Skyhawks. This marked the end of service of attack Skyhawks with the US Navy and the US Marine Corps. After that date, the only Skyhawks in Navy or Marine Corps service were two-seat trainers.

A low-vis clad A-4M from VMA-332 'Fighting Gamecocks' *CREDIT: US Navy*

An NATC high visibility A-4M *CREDIT: US Navy*

An OA-4M from MAG-32 at NA Cherry Point in May 1990
CREDIT: US Navy

An OA-4M from H&MS-32 in an experimental paint scheme
CREDIT: US Navy

OA-4M

The designation OA-4M was assigned to twenty-three TA-4Fs that were reworked at NAS Pensacola to act as fast forward air controller aircraft for the Marine Corps. Whilst this version should ostensibly be in the two-seat Skyhawk chapter, as it was an operational, rather than a training aircraft, it deserves it place here. The OA-4M was basically a TA-4F equipped with the electronics of the A-4M, the main US Marine Corps close support aircraft. The modifications included a ground control bombing system, a KY-28 secure voice system, a canted in-flight refuelling probe, extra cockpit side armour plating, an APN-194 altimeter, and ARC-159 radio and ARC-114 VHF radio. Perhaps the most visible change was the fitting of the dorsal electronics hump, neatly faired into the rear of the two-seat canopy. The nose sensor group of the OA-4M was basically the same as that of the A-4M, but the Angle/Rate Bombing system was not fitted, and the laser spotting and tracking window was faired over, since neither of these were needed on an aircraft which was not to be employed in a pure bombing role. Some consideration was given to fitting the Pratt & Whitney J52-P-408 of the A-4M, but this turned out to be impractical, and the OA-4M retained the J52-P-8 of the TA-4F. The first OA-4M, BuNo 165294 was tested at NATC Patuxent River, MD in July of 1978. The first squadron to receive the OA-4M was H&MS-52, based at MCAS Cherry Point.

An OA-4M with Harpoon missile fitted
CREDIT: US Navy

A Marines OA-4M seen at Utapo Air Base
CREDIT: US Navy

CHAPTER 3 Two Seat Skyhawks

A Fleet Composite Squadron (VC-10) TA-4J Skyhawk armed with Sidewinders and Mk.20 CBUs.
CREDIT: US Navy

Almost from the very beginning, Douglas had suggested to the Navy that a two-seat version of the Skyhawk would be advantageous, both as a trainer as well as for some types of combat missions where a second pair of eyes might be useful. These requests had always been turned down by the Navy on budgetary considerations. However, in 1964 the Navy changed its mind and convinced the Department of Defence to allocate funds for building of two prototype two-seat Skyhawks. One of the more convincing arguments for the Skyhawk trainer was that those single-seat Skyhawks, then being used for Stateside training, could be released for combat duty in Vietnam. The existing A-4E contract was amended, calling for the last two A-4Es, BuNos 152102 and 152103to be taken off the production line and modified into two-seaters under the designation TA-4E.

A TA-4J Skyhawk of Training Wing 3 (TW-3) waits behind the blast deflector for its turn at the catapult as another clears the flight deck of the auxiliary aircraft landing training ship USS Lexington during pilot carrier training in the Gulf of Mexico in early 1989
CREDIT: US Navy

TA-4E

The TA-4E retained the engine, armament, and most of the airframe of the A-4E. A twenty-eight-inch fuselage plug was added, that incorporated a second cockpit with a raised seat, and both cockpits were enclosed underneath a single large canopy that opened upward via a hinge at the rear. A set of dual controls were provided, and the two crew members sat on Escapac 1C-3 zero-zero rocket ejector seats. To allow for a second cockpit, and necessary controls and instrumentation, a twenty-eight inch 'plug' was constructed to lengthen the fuselage between the single seat cockpit and the fuselage fuel tank immediately behind it, and the fuselage fuel tank was reduced in size from 810 gallons to 670 gallons. Wing lift spoilers were installed above the flaps in the upper wing trailing edge, which were designed to improve crosswind landing performance, and the nose wheel steering was provided to allow for better control during crosswind taxiing and during ramp or carrier deck manoeuvring. The TA-4E was powered by a Pratt &

A colourful TA-4J from Fleet Composite Squadron 1 (VC-1)
CREDIT: US Navy

Whitney J52-P-8A engine, and TA-4E BuNo 152102 flew for the first time on June 30, 1965. Since the TA-4E incorporated many improvements over the A-4E, the designation was changed shortly thereafter to TA-4F to indicate that it was basically a new aircraft.

TA-4F

The TA-4F entered service in May 1966 with VA-125 at NAS Lemoore in California, which was an A-4 Combat Readiness Air Group, and retained all the combat capability of the single-seat Skyhawks. A total of 241 TA-4Fs were built, including the two aircraft originally ordered as TA-4Es. They were issued to several Navy and Marine Corps units, and most, if not all the Marine and Navy Skyhawk attack squadrons had at least one or two TA-4Fs assigned. One of the more useful roles of the TA-4F was as forward air controller for the Marine Corps. They helped to direct Marine pilots and aircraft in identifying and pinpointing targets in Vietnam. Twenty-three of these aircraft were later modified to the OA-4M configuration. Four aircraft were modified as EA-4F electronic warfare trainers. They were modified to carry external stores that could electronically simulate the signals of Soviet missiles and aircraft. The four planes were issued to VA-33. Many TA-4Fs were later converted to TA-4J configuration when they went through overhaul. Additionally, four TA-4Fs were converted for ECM training as EA-4Fs and assigned to VAQ-33

Four TA-4Fs were converted for ECM training as EA-4Fs and assigned to VAQ-33
CREDIT: US Navy

A TA-4F from Headquarters and Maintenance Squadron 32 (H&MS-32) preparing to taxi from the flight line at Marine Corps Air Station Cherry Point
CREDIT: US Navy

TA-4J

The TA-4J was a two-seat trainer version of the Skyhawk that was intended as a replacement for the Grumman TF-9J Cougar. The TA-4J was ordered in 1968 to provide advanced training for Navy and Marine Corps pilots. The TA-4J was basically identical in structure to the earlier TA-4F, but differed in having a lot of the tactical weapons systems removed. The ability to use the inflight buddy refuelling pod was deleted, but the mid-air refuelling probe was retained. The outer underwing pylons were also removed, reducing the number of pylons to three. The deletion of the offensive weapons capability resulted in a reduction in the empty weight of 230 pounds. A lower-rated J52-P-6 engine was installed,

A USMC TA-4F Skyhawk of H&MS-13 in Vietnam. H&MS-13 was based at Chu Lai, and used the TA-4Fs as forward air control aircraft. This Skyhawk is armed with Mk.82 bombs.
CREDIT: US Navy

A TA-4J Skyhawk VT-7 waits to be launched during flight operations aboard the USS Carl Vinson
CREDIT: US Navy

A TA-4F Skyhawk Marine Corps Headquarters and Maintenance Squadron H&MS-13 in flight on 15 July 1975
CREDIT: US Navy

but the twin 20mm cannon were retained, however, one or both guns were usually removed in the field. The TA-4Js were painted in a white with red trim painting scheme, which replaced the grey and white scheme of fleet Skyhawks, and the TA-4J flew for the first time on December 17, 1968, and entered service with VT-21 in mid 1969. The primary role of the TA-4J was to train carrier pilots, and a total of 226 TA-4Js were built. Many TA-4F airframes were converted to TA-4J configuration by removal of their offensive weapons systems. Several TA-4Js were used in support roles, including that of adversary aircraft.

Two TA-4Fs from VA-127 'Royal Blues' in flight
CREDIT: US Navy

A TA-4J from VT-7 prepares to launch from there USS Carl Vinson
CREDIT: US Navy

CHAPTER 4

Skyhawks in the Vietnam War

A-4F Skyhawks parked on the flight deck of the USS Hancock, and armed for a mission over Vietnam on 25 May 1972
CREDIT: US Navy

A-4 Skyhawks were the US Navy' and Marine Corps primary light attack aircraft used over North Vietnam during the early years of the Vietnam War, until later being supplanted by the A-7 Corsair in the light attack role. Skyhawks carried out some of the first air strikes by the US during the conflict, and a Marine Skyhawk is believed to have dropped the last American bombs on the country. Notable naval aviators who flew the Skyhawk included Lieutenant Commanders Everett Alvarez, Jr. and John McCain, and Commander James Stockdale. On 1 May 1967, an A-4C Skyhawk piloted by Lieutenant Commander Theodore R Swartz of VA-76 aboard the carrier USS Bon Homme Richard, shot down a North Vietnamese Air Force MiG-17 with an unguided Zuni rocket as the Skyhawk's only air-to-air victory of the Vietnam War! In August 1964, the 'Gulf of Tonkin' incident resulted in the escalation of US military efforts into what became known as the Vietnam War. By that point, the Navy had standardised on the A4D, as its light jet attack aircraft in carrier air wings, and proved to be an aircraft well able to punch above its weight. In early February 1965 after a Viet Cong attack on a US base at Pleiku, saw the US aircraft carriers on station launch major retaliatory strikes in Operation 'Flaming Dart'. These strikes were met with heavy anti-aircraft fire and did nothing except see the North Vietnamese, support the Viet Cong, blowing up the US troop barracks at Qui Nhon. The Navy responded and the air war over Vietnam began. The air strikes in February were followed on 2 March 1965 by Operation 'Rolling Thunder', which was intended to significantly disrupt the North Vietnamese ability to supply their forces and wage war in South Vietnam. This included bombing bridges, petroleum storage facilities, ammunition storage facilities, and supply trucks on the road. The concept was that the bomb line would gradually proceed north toward Hanoi with the expectation that the North Vietnamese would give up their support of the Viet Cong before then. 'Rolling Thunder' failed to meet those objectives and ended in October

An A-4E Skyhawk fromVA-106 'Gladiators' is brought to the launching position on a steam catapult of the USS Intrepid during flight operations in the Gulf of Tonkin, September 1968 *CREDIT: US Navy*

An A-4C Skyhawk of VA-34 'Blue Blasters' taxies into position for launching aboard USS Saratoga *CREDIT: US Navy*

1969. However, the conflict continued unabated.

The Skyhawk was not limited to carrier-based operations during the Vietnam War, and in March 1965, the Marines landed at Da Nang, South Vietnam, to protect the airfield there from Viet Cong attacks and establish an airbase for its aircraft that were providing air support and to shorten the flight time to targets in areas to the south. Da Nang aircraft assigned to Marine Air Group 11 provided missions to the north, over the DMZ, and into southern North Vietnam and MAG 12 was assigned to a new airfield for combat operations in South Vietnam. Chu Lai, a sandy beach about 50 miles south of Da Nang and on the coast, was selected as the base for MAG 12 and its A-4 Skyhawks, and here Navy Seabees initially created a 3,500-foot runway by laying down an aluminium plank surface, with arresting gear was then installed for landings. Because of the short runway, the A-4s used JATO bottles for take-offs from Chiu Lai, and a SATS (Short Airfield Tactical Support) catapult installation was also completed in May 1966 to replace the JATO take-offs. The A-4s could then be launched by a shuttle cart and steel cables attached to the nose wheel and driven by two General

An A-4C from VA-112 'Broncos' lands aboard the USS Kitty Hawk off Vietnam *CREDIT: US Navy*

An A-4C Skyhawk of VA-153 'Blue Tail Flies' aboard the USS Coral Sea during operations in the South China Sea, 24 March 1965 *CREDIT: US Navy*

An A-4C Skyhawk from VMA-225 'Vagabonds' makes the first landing at Chu Lai, Vietnam
CREDIT: US Navy

A pair of A-4C Skyhawks from VA-113 'Stingers 'on the catapult of the USS Kitty Hawk during that carrier's deployment to Vietnam from 19 October 1965 to 13 June 1966. The Navy experimented with aircraft camouflage and painted half of the aircraft of CVW-11 with dark green colours to blend in with the Vietnamese jungle. The results proved inconclusive for the. Navy, whereas the Air Force concluded that the camouflage was effective. Note that oil from the engine oil breather has streaked the camouflage paint
CREDIT: US Navy

Electric J79 jet engines driving geared turbines. The Skyhawk would be attached to the shuttle by cables fitted to the catapult hooks with its nose wheel resting on the shuttle. However, the original runway was supplemented with a parallel 10,000ft concrete runway in late 1966, ending the use of the SATS catapult and arresting gear. The JATO bottles were used to give the Skyhawks an extra push on take-off. These dry-fuel rockets were attached to the rear fuselage of the aircraft, giving five-second burn that generated 3,500 pounds of thrust and cut the take-off distance in half.

On 1 June 1965, the first VMA-225 A-4C Skyhawks landed, and undertook their first close air support mission that afternoon, whilst four more Skyhawks from VMA-311 arrived the next day. With the A-4Cs of VMA-225 and the A-4Es of VMA-311 in place at Chu Lai, MAG-12 added a third Skyhawk combat unit, VMA-214 Black Sheep, before the end of June 1965. Also, flying the A-4C came VMA-214, and MAG-12 increased its total A-4 in the autumn of 1965 with the arrival of VMA-211, the famed 'Wake Island Avengers'. Chu Lai's A-4 squadrons were regularly rotated back to Japan as others came in to fly combat tours. VMA-224 arrived in October, to be followed by VMA-223 in December. In addition to having considerable success with radar bombing, Marine A-4 pilots also worked with Air Force forward air controllers and used their

An A-4C Skyhawk of VA-113 'Stingers' landing aboard the aircraft carrier USS Kitty Hawk in the South China Sea after a strike on the Viet Cong in South Vietnam
CREDIT: US Navy

An A-4C Skyhawk from VA-172' Blue Bolts' is fully armed with bombs on the flight deck of the aircraft carrier USS Franklin D. Roosevelt off Vietnam, in 1966
CREDIT: US Navy

An A-4E Skyhawk of VA-23 'Black Knights' is launched from the USS Coral Sea for a mission in Vietnam. The aircraft is armed with AGM-45 Shrike missiles and LAU-3 rocket launchers *CREDIT: US Navy*

An A-4E Skyhawk from VA-23 'Black Knights' fires 70mm (2.75 in) rockets at Viet Cong targets in South Vietnam *CREDIT: US Navy*

own tactical air coordinator aircraft, which as of 1967 were two-seat TA-4Fs. Retaining the two-cannon armament and similar stores delivery capability of the single seaters, the TA-4F dual-control trainer also generally duplicated the combat models' avionics fit, making for ease of maintenance. Combat operations continued for the rest of the War until in MAG-12 and VMA-211 departed Vietnam in February 1970 and relocated in Japan, while VMA-223 returned to the US. VMA-311 moved to Da under the operational control of MAG-11 and continued to support the ongoing war in Laos and Cambodia. VMA-311 flew into Bien Hoa, near Saigon, and began flying combat missions into Cambodian border regions. After the January 1973 cease-fire, VMA-311 was to be the last air combat unit of all the US services to leave the country.

At the outset of the War the Skyhawks weapons involved their 20mm cannons, unguided rockets and iron bombs, with the AGM-12 Bullpup being the only guided weapon. Also available was the 2.75 FFAR (Folding Fin Aircraft Rocket) in 5 inch Zuni unguided rockets, both ideal for Flak suppression, and the

A-4E Skyhawks from VA-163 'Saints' and VA-164 'Ghost Riders' attacking the Phuong Dinh railroad bypass bridge, 10km north of Thanh Hoe, North Vietnam, on 10 September 1967 *CREDIT: US Navy*

A bomb-laden Marine Corps A-4E Skyhawk of VMA-121 'Green Knights' at Chu Lai, Vietnam *CREDIT: US Navy*

A Marine Corps A-4E Skyhawk of VMA-224 'Bengals' taking off on a combat mission from Chu Lai, Vietnam, on 24 September 1966 *CREDIT: US Navy*

An A-4F Skyhawk of VA-153 'Blue Tail Flies' is readied for launching from the USS Coral in September 1968 *CREDIT: US Navy*

An A-4E Skyhawk of VA-164 'Ghost Riders' en route to a target in North Vietnam armed with six Mk.82 bombs and two AGM-12 Bullpup missiles *CREDIT: US Navy*

An A-4F Skyhawk of VA-212 'Rampant Raiders' in flight during a mission over North Vietnam, in the summer of 1968 *CREDIT: US Navy*

Cluster Bomb unit (CBU) was and alternative to the cannon and FFAR. Snakeye retarded bombs were also used but generally released with the retard function disabled, and Marine Skyhawks made good use of the cannon/Snakeye combination in CAS missions. Due to the increasing US air power available in daylight hours on the supply routes, the North Vietnamese switched to travelling by night, and so the carrier strike aircraft would need to go nocturnal for their truck hunting missions. Although a handful of Skyhawks carried a small radar it was of little use for this type of mission so A-4 crews resorted to Korean War tactics. So, if they saw anything of interest they would drop magnesium flares and attack by their light. The Marine Corps also used TA-4Fs to mark targets with flares and rockets to allow F-4 Phantoms to then attack the targets. Walleye I glide bombs were first used in 11 March 1967 when A-4E's from VA-212 successfully attacked military targets. The next day Walleye was used to attack the Thanh Hoa Bridge, without success. The more powerful Walleye II were used in October and on March 1967, two were used to destroy what was left of the Thanh Hoa Bridge. Other innovation in ordnance was the so-called 'Bullwinder', a combination of Bullpup and Sidewinder components, and the 'Big Mutha' Built-In Guided M-118 Under Terminal Homing for Accuracy bomb. During the latter part of the War Laser Guided Bombs began appearing, however, as the A-4 was now in the process of being replaced aboard carriers, their use was limited. Some VA-164 A-4Fs were equipped with Laser Spot Trackers in 1973, fitted in the nose and displayed via a Ferranti gun sight. Also trialled was the Light Weight Laser, a small portable unit used aboard Marine Corps TA-4Fs.

Surface to Air Missiles were the bane of the US air forces during the war, and one of the pieces of equipment fitted to the Skyhawk was the ALQ-151 Defensive Electronic Countermeasures system. So, equipped aircraft from VA-153 and VA-155 had 'ice cream cone' antennae's, one each on the nose and tail and two on the undersides below the guns. This system was designed to spoof the Fan Song radar to believing the aircraft was in a slightly different position to its actual plot. The next system was designed to warn the pilot of a missile launch and allow him to

An A-4E from VA-106 'Gladiators' is positions aboard the USS Intrepid in June 1968 *CREDIT: US Navy*

An A-4E from VA-115 'Silver Foxes' gets a ';wave-off' during a deployment avoid the USS Constellation to Vietnam in December 1966 *CREDIT: US Navy*

time to initiate an escape manoeuvre. As such the AN/APR-27 Launch Warning Receiver was fielded in 1966. Further enhancements came in the shape of the AN/APR-25 Radar Homing and Warning System, and the ALE-29 dispenser added to either side of the lower fuselage. Another weapon in the SAM fight was the AGM-45 Shrike anti-radar missile, with VA-23 being deployed aboard the USS Midway with pre-production version and fired the first Shrike in combat on April 25, 1965. In addition, the SAM menace, MiGs were a constant threat, which pilots viewed with apprehension and anticipation, and Skyhawks were often attacked by MiGs, but their manoeuvrability made it possible for them to evade most of these attacks. Losses of Skyhawks to AAA and SAMs were heavy, especially during the initial stages of the Vietnam conflict. Initially, attacks were carried out at low level to avoid detection by radar, but it was soon discovered that the required zoom before bomb release resulted in an unacceptable loss of speed and manoeuvrability. A change of tactics was called for, and targets were later approached at high altitude and high speed, and were bombed in shallow diving attacks. A high-altitude approach opened-up a vulnerability to surface-to-air missiles, but the North Vietnamese SA-2 SAMs were relatively easy to avoid, provided that the pilot saw one coming. The Skyhawks days in Vietnam were being brought to a close when in late 1967, the first A-7 Corsair, the A-4s replacement arrived aboard the USS Ranger, and over the next few years almost all the Navy's light attack units would convert to the A-7, the Marine Corps however, continued to use the A-4 in combat.

An A-4E from VA-155 'Silver Foxes' is manoeuvred on the deck of the USS Coral Sea *CREDIT: US Navy*

CHAPTER 5 Skyhawk Aggressors

A pair of A-4E Skyhawks from VF-43
CREDIT: US Navy

The small size of the A-4 Skyhawk, coupled with its superb low-speed handling and agility, plus a well-trained pilot at its controls made it an ideal platform to teach fleet-flyers the finer points of DACT. To better perform the Adversary role, single- seat A-4E and F models were introduced to the 'Aggressor' role, as were TA-4J and TA-4F two-seat aircraft, but the ultimate 'bandit' Skyhawk had to be A-4F 'Super Fox', which was equipped with the uprated J52-P-408 engine, like the configuration used by the Blue Angels. And with these nimble performers mimicking the tactics of aircraft from around the world the adversary squadrons soon began to display vivid 'threat-type' paint schemes on their aircraft, signifying their primary role of Adversary training. A surplus of former USMC Skyhawk's resulted in a few A-4M versions being used

An A-4E from VC-5
CREDIT: US Navy

An A-4F of the Naval Strike and Air Warfare Center at NA Fallon
CREDIT: US Navy

by both VF-126 and 'Topgun' and even though the A-4 was augmented by the F-5E, F-21, F-16 and F/A-18, in the Adversary role, the Skyhawk remained a viable aerial threat until it was retired by VF-43 in 1993 and shortly thereafter by VFC-12, with the last A-4 fleet operators VC-8, relinquishing its Skyhawk's in 2003. With a renewed emphasis on Air Combat Manoeuvring training, brought about by the establishment of the Naval Fighter Weapons School – 'Topgun' in 1968, and the availability of A-4 Skyhawk's from both the Instrument RAGs and Composite Squadrons at the Master Jet Bases to provide a ready resource of lively fleet-footed aircraft, the powers that be at 'Topgun' decided to use these nimble aircraft as a 'surrogate' for

Bandit and Prey - An A-4F manoeuvres behind an F-14A
CREDIT: US Navy

A line up of VF-43's A-4Fs at NAS Oceana
CREDIT: US Navy

An A-4F from VF-43 'Challengers'
CREDIT: US Navy

the MiG-17 and MiG-21. At the time, the F-4 Phantom was just beginning to be exploited to its full potential as a fighter and to that point had not performed as well as expected against the smaller North Vietnamese 'Fresco' and 'Fishbed' opponents. Therefore 'Topgun' introduced the notion of Dissimilar Air Combat Training (DACT) using stripped down A-4's in the so- called 'Mongoose' configuration with fixed slats and basic avionics. The Skyhawk thus went on to serve with distinction with units such as VF-126 at NAS Miramar, VA-127 at NAS Lemoore, VF-43 at NAS Oceana and VA-45 at NAS Cecil field and additional single-seat A-4 Skyhawk's were also assigned to composite squadrons (VC) worldwide to provide training and other services to deployed units and these included VC-1 at NAS Barbers Point, VC-7 at NAS Miramar, VC-5 at NAS Cubi Point, VC-8 at NS Roosevelt Roads, VC- 10 at Guantanamo Bay and VC-12 and VC-13.

A pair of VC-5 A-4s get airborne from NAS Oceana
CREDIT: US Navy

The Naval Fighter Weapons School – 'Topgun' was originally established as a 'school' within VF-121, a F-4 Phantom Replacement Air Group (RAG) at NAS Miramar and tasked with providing graduate level Air Combat Manoeuvring training, with the first class commencinginstartedMarch1969. The'school'thenevolvedin January 1972 into a separate VF-121 detachment, all the while using VF-126 A-4's and maintenance support personnel. The NFWS instructors flew the A-4's, and each crew going through as students brought with them their own squadron maintenance personnel and own squadron aircraft. Thus, in July 1972 the NFWS was given full

Carrying an ACMI pod is the A-4F from VFC -13 'Saints'
CREDIT: US Navy

An A-4F from VA-127
CREDIT: US Navy

squadron status as the Naval Fighter Weapons School and handled the graduate courses for Navy pilots, while VF-126 was more of a day-to-day adversary unit, supporting the F-4's and F-14's at Miramar. When the Skyhawk was being replaced as the fleet attack aircraft in 1967, VF-126 was already providing aircraft for aggressor training and began taking on more A-4's for the 'bandit' mission aircraft because of its small size, manoeuvrability and smokeless engine, and the unit began providing dedicated adversary missions from NAS Miramar in April 1967 and continued until they were disestablished in April 1994.

VFA-127's history began at NAS Moffett Field, CA when it was part of VF-124, the F-8 RAG squadron. VA-127 was redesignated VFA-127 on March 1, 1987, in keeping with its new role as the Pacific Fleet adversary squadron responsible for training Naval Strike Fighter pilots in air-to-air combat and flying amongst other aircraft the A-4 Skyhawk. In October1987, the squadron moved to NAS Fallon, making VFA-127 the only Navy squadron permanently based in Nevada where they became known as the 'Desert Bogeys'. VFA-127 became the Navy's last active-duty adversary squadron at NAS Fallon and was finally disestablished on 23 March 1996, with their 'bandit' role being passed to the 'Fighting Saints' of Fighter Squadron Composite 13 and whilst VFA-127 flew A-4's, F-5E in the adversary role, now VFC-13 only flies F-5E's.

VC-8 'The Redtails' was the final squadron in the US Navy to fly the TA-4J Skyhawk and its mission included tracking exercises for the fine-tuning of ship's radar's, ACM for allied aircraft, and launching AQM and BQM aerial target drones. The squadron was deactivated on the 23 August 2003.

'Top Gun' A-4F at Miramar
CREDIT: US Navy

Colourful Aggressor Gallery

An A-4E of VF-127

A-4E,Red 02 of VF-127 'Cyclons'

A-4F 155018 of VF-126

A-4E 149456 of the NFWS

A-4E 149656 of the NFWS

A-4E 150009 of the NFWS

A-4E 150023 of the NFWS

A-4E 150090 of the NFWS

A-4E 155044 of the NFWS

TA-4J 153683 of the NFWS

TA-4J 154327 of VF-45

A-4E 149066 of VF-126

A-4E 150073 of VF-126

A-4F 154190 of VA-127

A-4E 151038 of VF-45

TA-4J 153690 of VF-127

TA-4J 158465 of VF-127

Skyhawk Walkaround

A-4F Skyhawk 155025 - VA-94 US Navy

PRESERVED AT NAS FALLON AIR PARK

Photographs by of Luc Colin

Tailhook fairing (port)

Tailhook fairing (starboard)

Vortex generators

Main wheel from the rear

Nose wheel

Nose wheel leg

Nose wheel door inner face

Main wheel inner face

Main wheel detail

Canopy and forward fuselage

Access panel

Main wheel door inner face

IFR probe – mid-section

IFR Probe profile

IFR probe front detail

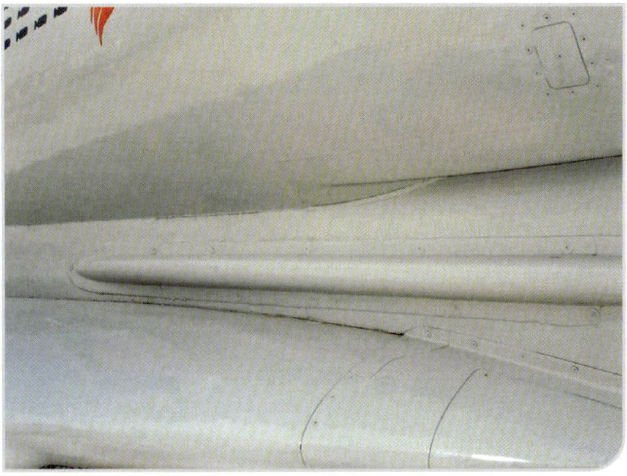
IFR probe fuselage join

Intake in detail

Rear fuselage

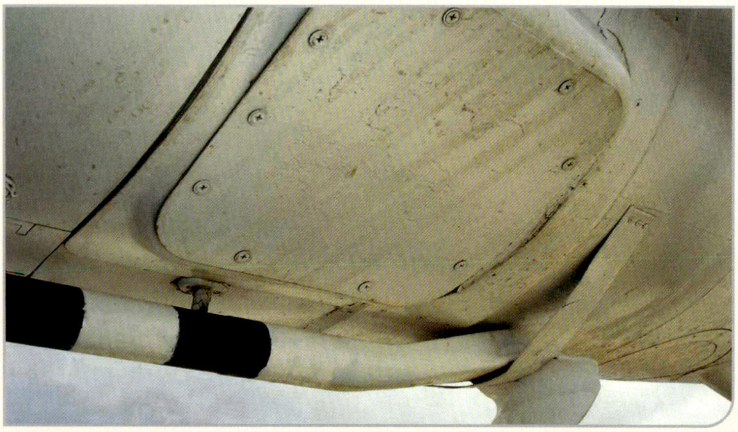

Arrestor hook

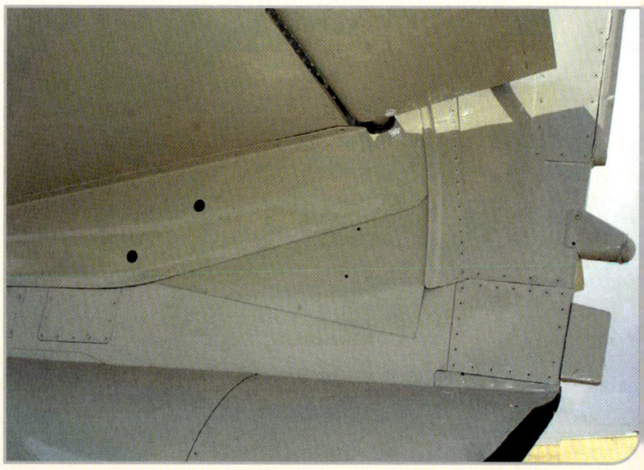

Under tail detail

Main wheel door profiles

Main Wheel door outer face

Inside main wheel bay

A-4M Skyhawk 160264 – US Marine Corps

PRESERVED FLYING LEATHERNECK MUSEUM MCAS MIRAMAR

Photographs by of Luc Colin

Forward fuselage

Tail detail

Rear fuselage detail

Pylon detail

Nose sensors

ARBS detail

Landing light

Main wheel bay detail

Centreline pylon detail

APPENDIX

I Specifications

A quartet of A-4Ms from VMA-214 'Black Sheep' *CREDIT: US Navy*

General Charcteristics

Crew: 1

Length: 40ft 1.5in (12.230m)

Wingspan: 27ft 6in (8.38m)

Height: 15ft 2in (4.62m)

Empty weight: 9,853lb (4,469kg)

Gross weight: 16,216lb (7,355kg)

Max take-off weight: 24,500 lb (11,113kg)

Powerplant: 1 × Pratt & Whitney J52-P-6A turbojet engine, 8,500lbf thrust

Performance

Maximum speed: 673mph, 1,083km/h at sea level

Range: 1,008nmi

Ferry range: 2,194nmi

g limits: +8 -3

Rate of climb: 5,750ft/min

Wing loading: 62.4lb/sq ft

Armament

Guns: 2× 20mm Colt Mk.12 cannons, 100 rounds/gun

Hardpoints: 4× under-wing & 1× under-fuselage pylon stations with a capacity of 8,500lb, with provisions to carry combinations of:

Rockets:
4× LAU-10 rocket pods (each with 4× 127mm Mk.32 Zuni rockets)

Bombs:
- 6× Rockeye-II Mark 20 Cluster Bomb Unit (CBU)
- 6× Rockeye Mark 7/APAM-59 CBU
- Mark 80 series of unguided bombs (including 3kg and 14kg practice bombs)
- B43 nuclear bomb
- B57 nuclear bomb
- B61 nuclear bomb

Other:
- up to 3× 370 US gallon Sargent Fletcher drop tanks for ferry flight/extended range/loitering time

Missiles:

Air-to-air missiles:
- 4× AIM-9 Sidewinder

Air-to-surface missiles:
- 2× AGM-12 Bullpup
- 2× AGM-45 Shrike anti-radiation missile
- 2× AGM-62 Walleye TV-guided glide bomb
- 2× AGM-65 Maverick

Avionics
Typical avionics fitted to A-4s

- Bendix AN/APN-141 Low altitude radar altimeter (refitted to C and E, standard in the F)
- Stewart-Warner AN/APQ-145 Mapping and Ranging radar (mounted on A-4F, also found on A-4E)

APPENDIX

II A-4 US Navy and Marine Corps Skyhawk Variants

An A-4E Skyhawk from VA-45 Det 1 'Blackbirds' aboard the USS Intrepid *CREDIT: US Navy*

XA4D-1 - Prototype (pre-1962 designation)

YA4D-1 - (YA-4A, later A-4A): Flight test prototypes and pre-production aircraft (pre-1962 designation).

A4D-1 - See A-4A (pre-1962 designation)

A4D-2 - See A-4B (pre-1962 designation)

A4D-2N - See A-4C (pre-1962 designation)

A4D-3 – Proposed advanced avionics version, none built (pre-1962 designation).

A4D-4 - Long-range version with new wings, cancelled.

A4D-6 - Proposed version, none built. (pre-1962 designation)

A-4A - Initial production version, 166 built

A-4B - Strengthened aircraft and added air-to-air refuelling capabilities, improved navigation and flight control systems, provision for AGM-12 Bullpup missile, 542 built.

A-4C - Night/adverse weather version of A4D-2, with AN/APG-53A radar, autopilot, LABS low-altitude bombing system. Wright J65-W-20 engine with 8,200 lbf (36 kN) of take-off thrust, 638 built.

A-4D - Designation not used to avoid confusion with the pre-1962 designation A4D.

A-4E - Major upgrade, including new Pratt & Whitney J52-P-6A engine, strengthened airframe with two more weapon pylons (for a total of five), improved avionics, with TACAN, Doppler navigation radar, radar altimeter, toss-bombing computer, and AJB-3A low-altitude bombing system. Many later upgraded with J52-P-8, 499 built.

An A4D-2 Skyhawk from VA-34 'Blue Blasters' in flight
CREDIT: US Navy

A-4F - Refinement of A-4E with extra avionics housed in a hump on the fuselage spine (this feature later retrofitted to A-4Es and some A-4Cs), wing-top spoilers to reduce landing roll out, nose wheel steering, and more powerful J52-P-8A engine, later upgraded in service to J52-P-408, 147 built. Some served with Blue Angels acrobatic team from 1973 to 1986.

A-4M - Dedicated Marine version with improved avionics and more powerful J52-P-408a engine, enlarged cockpit, IFF system. Later fitted with Hughes AN/ASB-19 Angle Rate Bombing System (ARBS) with TV and laser spot tracker, 158 built.

TA-4F - Conversion trainer – standard A-4F with extra seat for an instructor, 241 built.

TA-4J - Dedicated trainer version based on A-4F, but lacking weapons systems, and with down-rated engine, 277 built new, and most TA-4Fs were later converted to this configuration.

TA-4E - Two A-4Es modified as prototypes of a trainer version.

EA-4F - Four TA-4Fs converted for ECM training.

A-4L - 100 A-4Cs remanufactured for Marine Corps Reserves and Navy Reserve squadrons. Fitted with A-4F avionics including the fuselage 'hump' but retaining J-65 engine and three-pylon wing.

OA-4M – Twenty-three TA-4Fs modified for Forward Air Control duties for the Marines in Vietnam.

Colour Side Profiles

By Ryan Dorling

Douglas A-4B Skyhawk 145013/AK of VA-15 'Valions'

Douglas A-4F Skyhawk 154190/NF of VA-22 'Fighting Redcocks'

Douglas A-4E Skyhawk 152072 NAF China Lake

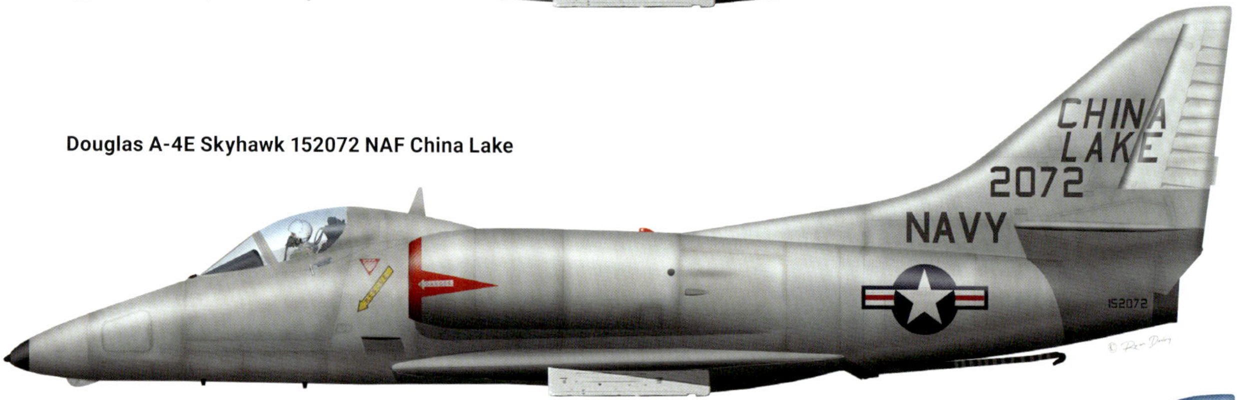

Douglas A-4E Skyhawk 150083/NL of VA-153 'Blue Tail Flies'

Douglas TA-4J 158147/N 0f VT-4

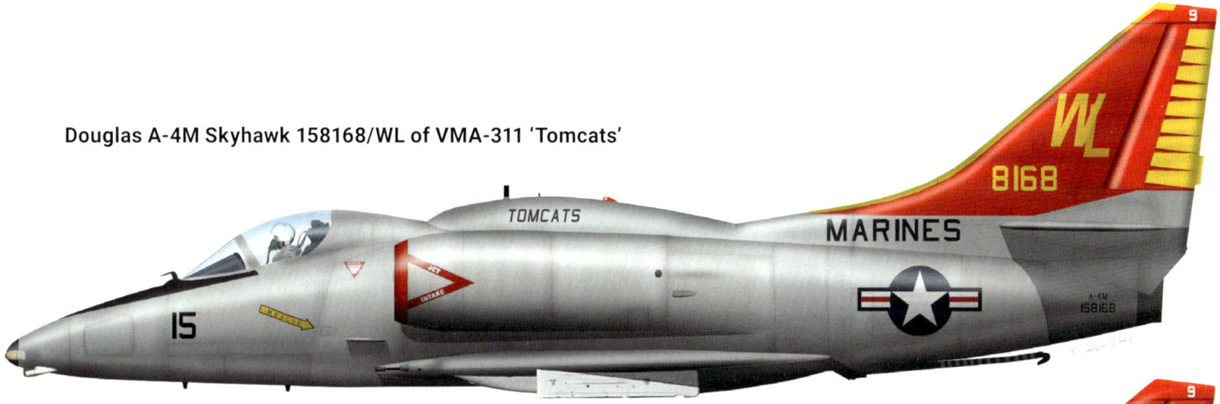

Douglas A-4M Skyhawk 158168/WL of VMA-311 'Tomcats'

Douglas A-4M Skyhawk 158192/SC of VMAT-102 'Skyhawks'

Douglas TA-4F Skyhawk 154311/EX of H&MS31

Douglas A-4M Skyhawk 160024/QG of VMFA-311 'Tomcats'

Douglas A-4E Skyhawk 'Aggressor' 149977/AD

Douglas TA-4J Skyhawk 'Aggressor' 154656/NJ of VFA-127

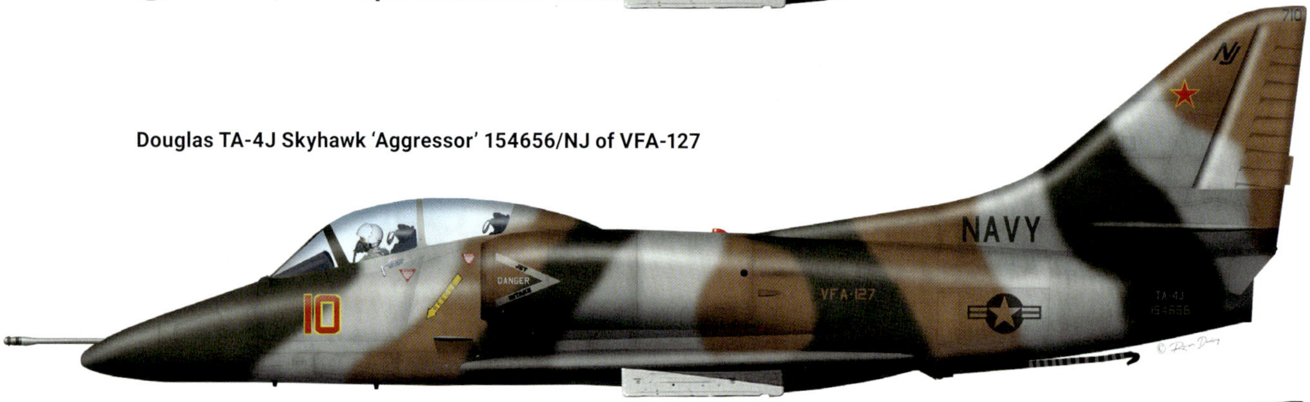

Douglas A-4F Skyhawk 155018/NJ624 'Aggressor'

Douglas A-4E Skyhawk 150023/56 'Aggressor'

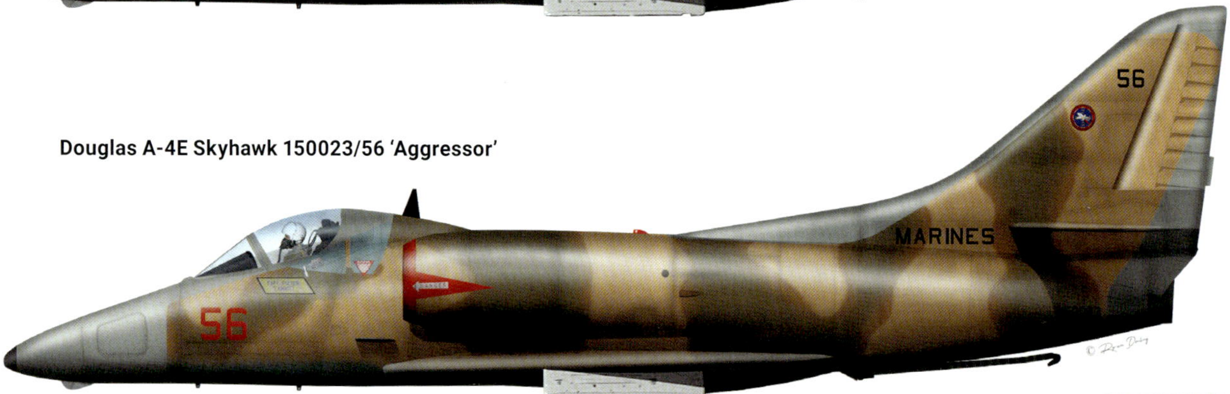

Douglas A-4M Skyhawk 159486/NJ632 of VF-126 'Bandits'

REAL TO REPLICA